My Favorite COLOR
ORANGE

A Crabtree Roots Book

AMY CULLIFORD

CRABTREE Publishing Company
www.crabtreebooks.com

School-to-Home Support for Caregivers and Teachers

This book helps children grow by letting them practice reading. Here are a few guiding questions to help the reader with building his or her comprehension skills.Possible answers appear here in red.

Before Reading:
- What do I think this book is about?
  - *This book is about the color orange.*
  - *This book is about things that are orange.*

- What do I want to learn about this topic?
  - *I want to learn what animals are orange.*
  - *I want to learn about shades of orang.*

During Reading:
- I wonder why...
  - *I wonder why some fish are orange.*
  - *I wonder why carrots are orange.*

- What have I learned so far?
  - *I have learned that tigers are orange.*
  - *I have learned that pumpkins are orange.*

After Reading:
- What details did I learn about this topic?
  - *I have learned that there are many shades of orange.*
  - *I have learned that food can be orange.*

- Read the book again and look for the vocabulary words.
  - *I see the word **pumpkin** on page 8 and the word **tiger** on page 6. The other vocabulary words are found on page 14.*

I see orange.

I see an orange **carrot**.

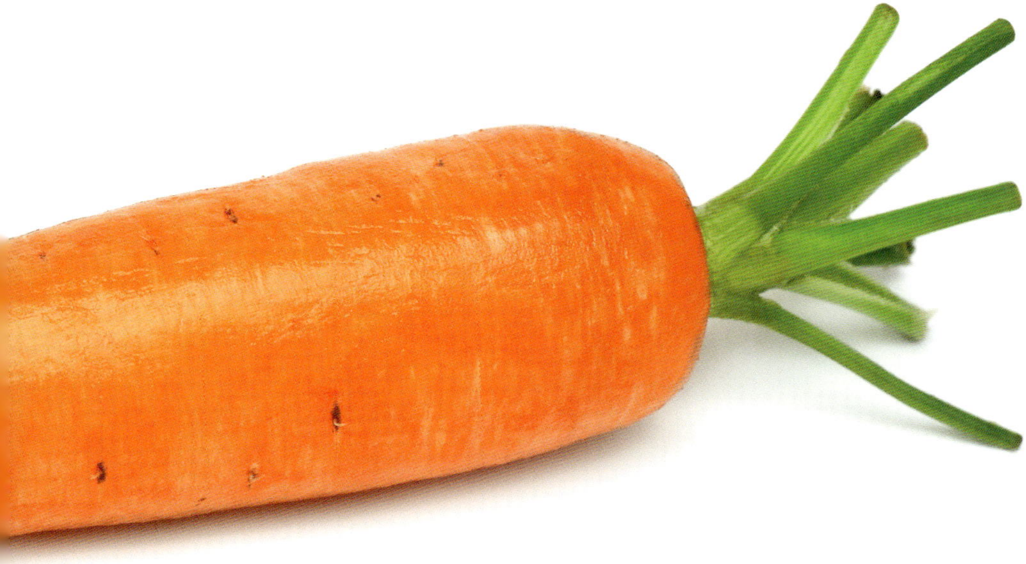

I see an orange **tiger**.

7

I see an orange **pumpkin**.

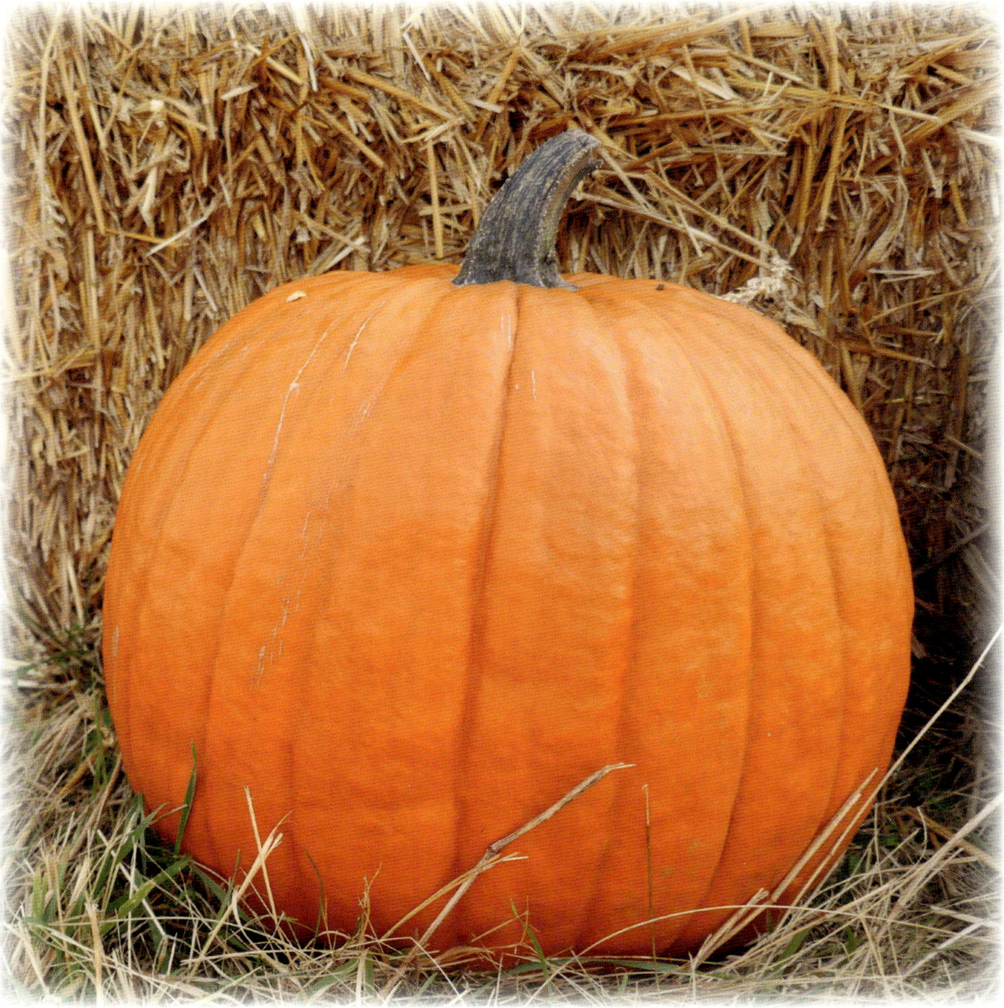

10

I see an orange fish.

What do you see that is orange?

13

Word List

Sight Words

an	is	what
do	orange	you
I	see	

Words to Know

carrot

fish

pumpkin

tiger

30 Words

I see orange.

I see an orange **carrot**.

I see an orange **tiger**.

I see an orange **pumpkin**.

I see an orange **fish**.

What do you see that is orange?

My Favorite COLOR

ORANGE

Written by: Amy Culliford
Designed by: Rhea Wallace
Series Development: James Earley
Proofreader: Kathy Middleton
Educational Consultant:
   Christina Lemke M.Ed.

Photographs:
Shutterstock: Africa Studio: cover; Iryna DenySova: p. 1; FeelIFree: p. 3; Flower Studio: p. 4-5, 14; Ondrej Prosicky: p. 7, 14; yul38885: p. 9, 14; Oleg_P: p. 10, 14; Drmitrystock: p. 13

Library and Archives Canada Cataloguing in Publication
Title: Orange / Amy Culliford.
Names: Culliford, Amy, 1992- author.
Description: Series statement: My favorite color | "A Crabtree roots book".
Identifiers: Canadiana (print) 20200384139 |
   Canadiana (ebook) 20200384147 |
   ISBN 9781427134684 (hardcover) |
   ISBN 9781427132598 (softcover) |
   ISBN 9781427132659 (HTML)
Subjects: LCSH: Orange (Color)—Juvenile literature.
Classification: LCC QC495.5 .C853 2021 | DDC j535.6—dc23

Library of Congress Cataloging-in-Publication Data
Title: Orange / Amy Culliford.
Description: New York, NY : Crabtree Publishing Company, [2021] |
   Series: My favorite color ; a Crabtree roots book | Includes index.
Identifiers: LCCN 2020050187 (print) |
   LCCN 2020050188 (ebook) |
   ISBN 9781427134684 (hardcover) |
   ISBN 9781427132598 (paperback) |
   ISBN 9781427132659 (ebook)
Subjects: LCSH: Orange--Juvenile literature. | Colors--Juvenile literature.
Classification: LCC QC495.5 .C855 2021 (print) | LCC QC495.5 (ebook) |
   DDC 535.6--dc23
LC record available at https://lccn.loc.gov/2020050187
LC ebook record available at https://lccn.loc.gov/2020050188

Crabtree Publishing Company
www.crabtreebooks.com   1-800-387-7650

Printed in the U.S.A./022021/CG20201130

Copyright © 2021 **CRABTREE PUBLISHING COMPANY**

All rights reserved. No part of this publication may be reproduced, stored in a retrieval system or be transmitted in any form or by any means, electronic, mechanical, photocopying, recording, or otherwise, without the prior written permission of Crabtree Publishing Company. In Canada: We acknowledge the financial support of the Government of Canada through the Canada Book Fund for our publishing activities.

Published in the United States
Crabtree Publishing
347 Fifth Avenue, Suite 1402-145
New York, NY, 10016

Published in Canada
Crabtree Publishing
616 Welland Ave.
St. Catharines, Ontario L2M 5V6